Hans Wehrli

An Axiom of Chirality as the Basic Principle of Physics

Hans Wehrli

An Axiom of Chirality as the Basic Principle of Physics

wbg Academic

Die Deutsche Nationalbibliothek verzeichnet diese Publikation
in der Deutschen Nationalbibliographie; detaillierte bibliographische
Daten sind im Internet über http://dnd.d-nb.de abrufbar

wbg Academic ist ein Imprint der Verlag Herder GmbH

Satz und eBook: Satzweiss.com Print, Web, Software GmbH
Gedruckt auf säurefreiem und
alterungsbeständigem Papier
Printed in Germany

Besuchen Sie uns im Internet: www.wbg-wissenverbindet.de

ISBN 978-3-534-40243-4

Elektronisch sind folgende Ausgaben erhältlich:
eBook (PDF): 978-3-534-40245-8
eBook (epub): 978-3-534-40244-1

Contents

1 Philosophy of Physics

The century-long, vain search for a comprehensive physical theory which makes the theory of relativity compatible with quantum theory and is able to describe all known interactions compels us to look for fundamentally new ideas. Fundamentally new ideas in a science always call for changes in the corresponding metascience. In the case of physics, that is metaphysics in its narrower sense, i.e. the science of language and methods of physics. In an on-going work, therefore, the basic metaphysical conditions – which are to comply with a physical theory – are examined, so that nature can be described as close to reality as possible. To do this, a dialogue between physics and philosophy is required. Most physicists underestimate the effect of their own epistemological prejudices on their research [Weinberg (1999); Rovelli (2010), 415]. Today natural laws are usually found by looking for experimental regularities with empirical perceptions of nature, whereas the metaphysical preconditions for the perceptions are ignored. With chirality theory, the laws of nature are as they must be for perception to be possible. The laws are not derived from nature, but as with Immanuel Kant [Kant (1783/2001)] and C.F. von Weizsäcker [Weizsäcker (1999)] from the way nature can be perceived, that is, from metaphysics in a narrower sense. This is the first paradigm shift of the new chirality theory.

It cannot be a task of this article to construct an axiomatic system for chirality theory. A mathematician once told me that for this goal 200 mathematicians and physicists would have to work for 20 years. But in a rather philosophical article like this it is necessary to list the contradictions between the usual axioms and physical perception. This concerns mainly the axioms of infinity and identity. The first is the main cause of the incompatibility of relativity theory with quantum theory, the latter for the neglecting of chirality as the basic principle of order in physics. However, there is no need to build up a complete axiomatic for chirality theory to understand the main ideas of its strategy.

In order to avoid misunderstandings, a few linguistic questions should be clarified from the outset. Languages have their limitations, which cannot in principle be overcome. Among other things, the meaning of the words or symbols is always vague, because for their definition one needs further words, which for their part, need to be defined in turn. If, for example, one defines space as the sum of all points and the point as an infinitely small place in space, then the definition turns in a circle. Such cycles are fundamentally inevitable, even if every child has a certain conception of what a point is and what space is. Nevertheless, there are a few important terms, as they are to be understood in this work, which should be defined.

Physics describes observed nature and formulates by means of mathematical methods laws of nature, which permit predictions (about the future). *Nature* is the entirety of all those things which can, in principle, be empirically – directly or indirectly – perceived [Whitehead (1939)]. In this sense, nature is *real* or *material*. *Existence* in chirality theory is an ontological property of anything that is either real or – if not real – must have a Be-ing in nature. For example, a point or a natural number is not real but it exists. A boson is directly, a fermion indirectly perceivable; therefore, both are real. Also black holes can be indirectly observed and are real. Space and time *per se*, all numbers except the natural numbers, and infinity are neither existing nor real but only ideas or mathematical models. The examples mentioned for this definition of reality and existence will be discussed later in the article. I am aware that depending on the philosophical and epistemological point of view there might be other definitions. *Mathematics* studies patterns in abstraction of the individual things which are patterned [Hampe (1998)]. *Empirical perception* is a flow of information from the outside into the conscious mind of a subject. The *subject* is an entity, which can take up, store and consciously process information. If the observer is transcendental, he is called a *presumed observer* and he is not part of reality. It is left open as to whether a subject can itself be part of nature. For the considerations in the present work it is sufficient to proceed from only one subject, the *ego* [Descartes (1641/1960)]. Possible further subjects and intersubjective communication are not brought up for discussion, because such further subjects can never be unequivocally differentiated from objects. An *object* is a summary of mathematical quantities or patterns, whose current values permit predictions about these very values (in the future) [Drieschner (1981)]. *Information* can be defined as answers to potential questions which can be reduced to a countable number of so-called *binary choices*, i.e. to alternatives which can be decided with a simple yes/no answer. The binary choices are computable as bits or qubits [Weizsäcker (1985), 163–173].

We shall now examine the extent to which today's usual methods and terms of physics contradict the definitions above. Subsequently, those basic metaphysical conditions for a physics which conforms better to these definitions are to be described.

2 Contradictions between Physics and Mathematical Models of Physics

2.1 Theory and Reality

Both the observing physicist and the thinking mathematician are real. Their observations and thoughts obey natural laws. Today most physicists are aware of that fact but in classical mechanics this has not yet been the case. The mathematicians even today seldom think about the natural limits of their thinking although there is no thinking without order, no order without chirality and no infinity of thinking. The question to the extent the subject influences the object by his thinking and observing is a very difficult one. Kant [Kant (1786/1957)] was convinced that even the thinking subject alters the object of his thinking by his thoughts. Thus, Kant might be considered the first quantum theorist. In quantum physics the interaction between observer and object is the topic of innumerable discussions and philosophical interpretations such as the Copenhagen interpretation.

The relation between observer and observed object is fundamental to any metaphysical strategy. It must be decided whether one wants to start with a transcendental subject or with a real observer. At first sight the real observer seems to be preferable. But in that case one has to describe a system consisting of both the observer as well as the observed object and of the relation between both. This system must be observed by another subject. If this also has to be real, then the system of the second observer who observes the observing first observer and the observed object has to be described and so on. In any case a presumed transcendental observer will have to be part of the metaphysical model. Therefore, I prefer to start with the simplest metaphysical strategy: That is a presumed transcendental observer who observes the real world and describes it by a mathematical model.

Later, in a second step, the presumed observer might be replaced by a real one. That should not require much adaption because the real observer has to obey the same natural laws that have been described with the presumed observer. The only new aspect will be the consciousness of the real observer, what might be philosophically rather pretentious [Blackmore (2003); Hofstadter (1985)]. Also the event counting presumed observer is "built into the model" but he is transcendental and therefore does not require an ambient space. He does not influence the observed object. I assume that mathematical models without such transcendental aspects are impossible for metaphysical reasons. This is often ignored by theoretical physicists who believe that they describe real nature as it is. As soon as they describe a real observer they introduce a presumed observer who observes the real one.

The physicist seeks to describe nature with the help of a set of mathematical tools. In so doing, he must always remain conscious of what in his science is mathematics and what is physics, because physics has the task of describing the reality of nature as authentically as possible, whereas mathematics, with its logic and axiomatic theory is actually a purely theoretical construct, which remains to a large extent independent of nature. In order to make by means of the laws of nature statements which are as reliable as possible, it is therefore important to be conscious of what in these laws is based on theory and what on reality. This consciousness has been increasingly lost in modern physics. Often there are ontological misunderstandings because it is not made clear whether a proposition concerns nature, the perception of nature or the mathematical model which describes that perception. E.g. the dimensionality might be 10 for nature, 4 for the perception of nature and infinite with the mathematical model that describes that perception (see Section 4.3). And possibly a new definition of the mathematical notion of dimensionality or of physical terms might be useful.

2.2 Simplicity as a Metaphysical Criterion

With respect to mathematics, the physicist is free to a large extent. There are, however, practical reasons to prefer one type of mathematics to another: First, mathematics should be simple, Dirac would say beautiful. I want a model that is mathematically as simple as possible. Simplicity is one of the metaphysical prerequisites of the model. This was emphasized by Aristotle, Kepler, Newton, Mach, Einstein, Heisenberg and Weizsäcker [Wehrli (2008), 34]. Therefore, the mathematical model of chirality theory uses as few elements as possible, namely points which have no other intrinsic property than their existence and no external properties except their relations to other points, i.e. to the structure of the model. Then I ask, whether the model can be physically interpreted. Second, the mathematics should not contradict empirical observation. Third, its logic should be convincing for the physicist and its axiomatic theory immediately evident. In mathematics an infinite number of logical and axiomatic systems is possible [Quine (1963)]. According to the above criteria the best mathematical model must be chosen. In quantum mechanics for example, the logical proposition $AB = BA$ contradicts empirical observation and therefore it has been dropped [Finkelstein (1996), 3ff.].

2.3 $A \equiv A$?

Similarly, the identity axiom $A \equiv A$ is questionable in physics: If the A on the left and the A on the right hand side of the identity sign were really identical, then one could not distinguish them. Since one can, however, always distinguish the two A's, this important proposition of logic is to be dispensed with if possible. The meaning of the A may remain open in this argumentation because two separate things – be it particles, observations, relations or times – *never* can be physically identical. Two distinct A's may have some equal properties which allow the physicist to define precisely the extent to which the A's are alike and the extent to which they are different. But in physics there are never "two things distinct" the observations of which are identical. There are only single distinct physical events. I admit that as soon as I introduce number theory and counting into my model, I have to accept the axioms of set theory and the identity axiom: $2 \equiv 2$ remains a true proposition. But then set theory is neither an aspect of nature nor one of the perception of nature, but only an aspect of the mathematical model which tries to describe that perception.

This is the second paradigm shift of chirality theory where the axiom $A \equiv A$ is dropped and arrows are introduced instead. Arrows have a direction and they connect a source and an end, i.e. two "things" that can be distinguished. It even seems to be possible to dispense altogether with the objects and deal only with arrows [MacLane (1971), p. 9]. This is the case with the physical axiom of chirality for the 4-point space (see Section 3.3). So arrows are suitable models of physical perceptions which are ordered, causal and take place in temporal processes. The mathematics to formulate such models probably is a finite category theory (see Sections 3. and 6.3) [MacLane (1971)]. Category theory is a mathematical alternative to set theory. Not the elements and their relations are the basis of category theory but morphisms and relations between morphisms. Both category theory and set theory are complementary.

Also in category theory, there is an identity axiom: For each object *A* an identity morphism exists [Dümont (1999)]. But unlike sets which obey the Leibniz principle of identity objects, category theory does not, i.e. indiscernibility does not necessarily imply identity [McLarty (1992)]. The mathematical axiom of identity probably has no physical meaning because in physics only events are observable. Events are the morphisms of physics which are to be mathematically described by category theory. Since with each event the structure of the universe is changed and is no more identical with the structure before the event there is no physical situation where the identity axiom is applicable (see Section 3.3). Also elementary particles are not identical in chirality theory, not even with themselves. Elementary particles exist as a process of events which indeed does not change their internal structure but always changes their relation to the rest of the universe. To ignore that relation is an artificial, theoretical and unphysical standpoint. Of course, for the description of particles it might be suitable to ignore their relation to the rest of the universe. E.g. a neutrino can be described ignoring the gravitational field it is the source of. This way the description might become mathematically identical with that of other neutrinos. But such a theoretical separation of the neutrino from the rest of the universe is always an artificial one since in chirality theory the relation between the neutrino and the universe changes with each event. The only observable property of the neutrino is the process of events and in nature there are never two distinct events which are identical (see Sections 4.2 and 4.9.1).

2.4 Infinity

Since infinity is never perceptible, not even – as sometimes is claimed – an "approximate infinity" a physical theory should completely do without the term infinity and thus without "higher mathematics" [Hilbert (1926); Weizsäcker (1971 a)]. Thus, irrational numbers (surds) such as π, e, $\sqrt{2}$ e.g., or integrating and differentiating are dropped and mathematics becomes substantially simpler without the axiom of infinity. The axiom of infinity by Ernst Zermelo reads: "There is at least one set within the domain which contains the empty set and one which contains the set a, whenever it contains a itself [Zermelo (1908)]." This contradicts physical perception because there are no such empty sets in physics. Events cannot be empty. They happen or they don't happen. Yes or no. The physicist as a real observer counts the events and the counted numbers always are finite. It is always clear where one event ends and where a new event begins and the counting is done by natural numbers (see Section 3.3). π, e and $\sqrt{2}$ are not aspects of nature.

The axiom of infinity changes mathematics considerably. As every axiom, it leads to less freedom for the mathematical model which is intended to describe physical perception. The axiom is not a prerequisite for all types of category theory. Patrick Suppes has shown that dispensing with the axiom of infinity has a similar effect to dispensing with the parallel postulate in Euclidean geometry. So this opens up similar perspectives to the introduction of curved spaces in the general theory of relativity [Suppes (1972)]. The renunciation of infinity and continuum is the third paradigm shift required by chirality theory.

2.5 Continuum, (Path-)Connectedness and Compactness

The renunciation of infinity in chirality theory means that no continuum is allowed, be it time or space of the model. A continuum is anything that goes through a gradual transition from one condition to a different condition, without any abrupt changes. In set theory the continuum is the real line, i.e. the line whose points are the real numbers $\mathbb{R}$. In the mathematical field of point-set topology, a continuum is a nonempty compact connected metric space. A topological space is said to be connected if it is not the union of two disjoint nonempty open sets. A set is open if it contains no point lying on its boundary. The space does not have to be Euclidean. In general topology, compactness is a property that generalizes the notion of a subset of space being closed (that is, containing all its limit points) and bounded (that is, having all its points lie within some fixed distance of each other).

There may be different notions of connectedness that are intuitively similar, but different as formally defined concepts. We might wish to call a topological space connected if each pair of points in it is joined by a path. However, this concept turns out to be different from standard topological connectedness; in particular, there are connected topological spaces for which this property does not hold. Because of this, a different terminology is used: spaces with this property are said to be path connected. While not all connected spaces are path connected, all path connected spaces are connected.

Now, the point is that the physical universe is finite by definition because otherwise it would in principle not be perceivable by a finite observer during his finite life time. The counting of events must be finite. Therefore, physical measurements always are a finite counting of events (see Section 2.8). The counting is done by finite natural numbers $\mathbb{N}$ and not by real numbers $\mathbb{R}$. In $\mathbb{R}$ there is a gradual transition from one number to the next without any abrupt changes, in $\mathbb{N}$ there is not.

According to chirality theory there is neither a "gradual transition" from one event to the next nor a path between the structures of the observed object before and that after the event (see Section 3.5, Figure 3). Therefore, it makes sense and simplifies the mathematical model if it is finite too. This excludes continua from the chirality model.

If this can be accomplished, the artificial renormalization and perturbation techniques can be dropped although they might still be useful to simplify certain calculations. Richard Feynman emphasized that with such techniques he can describe nature mathematically but he did not understand it because nobody can understand nature [Feynman (1985); Feynman (1966 a)].

Following all these paradigm shifts, not a lot remains of the mathematics we learned in school. For this reason, it is not yet possible to formulate the chirality theory in a precise mathematical language as it is usually done with the description of physical theories. The formalisms of quantum and general relativity theory as well as that of set theory cannot simply be applied, even if modified. Chirality theory is at a stage similar to that of classical mechanics in 1715 when Newton and Leibniz discussed the reality of space [Leibniz and Clarke (1715-16)]. Independently of one another they both discovered infinitesimal theory.

2.6 Do Space, Time and Photons exist?

Important terms of physics are also questionable because they describe things which cannot be observed empirically at all. This applies in particular to space and time, whereas the photon is not *per se* observed but rather its destruction. A space without objects and a time without events are inherently never perceivable. So space and time are not aspects of reality, but at most, questionable mathematical models in which we think [Einstein (1979)]. Instead, meaningful mathematical definitions for the terms object and event are required. The corresponding mathematical model then should show how space and time can be interpreted within that model. As yet, this has only been unsatisfactorily achieved in theoretical physics. In general relativity theory e.g. an event is simply a point in spacetime.

All sensory perceptions are conveyed exclusively by photons. These are thus central for every physical theory, which should conform with perception. Photons move at light speed, so their "internal clock" stands still according to the special theory of relativity. From the perspective of the photon it exists for zero seconds, even if from the physicist's perspective it has been travelling for billions of years. As the double-slit experiment demonstrates, the photon "is" during its existence not in a location, but in several locations simultaneously. When it meets our eye, it is completely destroyed; so strictly speaking, we do not see the photon, but the destruction of the photon. It is not surprising that a particle like the photon, which in certain respects exists for zero seconds, which is in no specific location and which cannot actually be perceived at all, is scarcely suitable as a model for the description of reality. This compels the metaphysician to ponder the meaning of the terms "is", existence and Be-ing.

2.7 Chirality as the Basic Principle of Physics

Physics proceeds from perception, i.e. from an information flow. The information flow is ordered in the direction from the object to the subject, otherwise it would become chaos. In order to define an order, one needs a duality of two distinguishable orientations. Every type of physical order is based on such a duality, be it yes/no-answers, left-handed/right-handed spin, past/future or positive/negative charge.

The duality of yes/no answers leads to an information theory which defines information as answers to potential questions which can be reduced to a countable number of so-called binary choices or bits, i.e. to alternatives which can be decided by a simple yes/no answer. The order or entropy of a system can be calculated from the probability of any answers to single questions [Weizsäcker (1985), 163–173].

The other types of duality, spin, time and charge are mathematically represented by the signature plus or minus or by the symbols $<$ and $>$.

Both types of order are based on chirality and they can be combined by means of the bit formalism. The imaginary qubit is a unit of quantum information – the quantum analogue of the classical bit. A qubit is a two-state quantum-mechanical system described by imaginary numbers. A metaphysical prerequisite of qubits is chirality because the chiral orientation of imaginary numbers must be defined. Therefore, chirality is more basic than information and the latter should be derived from the former. The mathematical advantage of qubit theory is that quantum theory is easily derivable from it. But since the quantum of qubit theory is not introduced as a metaphysical, but simply as a physical axiom, the qubit is no metaphysical help for understanding quantum theory.

In physics the two required chiral orientations are e.g. in time before and afterwards, in one-dimensional space left and right, in two-dimensional space clockwise and counter-clockwise, and with electrical charge positive and negative. Chirality (Greek *cheir* = hand) means handedness. The right hand is the mirror image of the left. Although both hands are isometric, they cannot be brought to coincidence with one another. So they are different from each other. However, a third hand, which is likewise isometric to the right and to the left hand, which nevertheless cannot be brought to coincide with either of them, does not exist. For each hand, there is one, and only one, counterpart with opposite handedness [Kant (1768/196)]. An object is called chiral if it has a mirror image which is not identical to the object.[1] Such a duality can be defined for mathematical or physical spaces with as many dimensions as desired. Every chiral object has an orientation. Its mirror image has the opposite orientation. Orientation is not an absolute characteristic; it only can be defined relative to the orientation of another, likewise chiral object or subject.

Chirality plays a fundamental role both in nature and in the metaphysics, because without this duality of Be-ing there is no order, no transmission of information and thus no perception, no memory and no thinking, not even on the assumption that thinking is something transcendental. In nature, one finds chirality from the most enormous galaxies through to living and dead matter and down to the smallest elementary particle, to the neutrino with its spin. The image and mirror image of space, time, spin and electrical charge have inherently different physical properties, an example being the weak interaction [Wolschin (1999)]. However, the *CPT* theorem (*CPT* symmetry) nevertheless shows that a connection between the orientation in space, time and the sign of the electrical charge must exist. Only if the orientations of space, time and electrical charge are reversed together the physical laws are preserved. This corresponds to the situation in category theory

---

1 "I call any geometric figure, or group of points, chiral, and say that it has chirality, if its image in a plain mirror, ideally realized, cannot be brought to coincidence with itself." [Boyle (1904)]

where the binary counterpart to each mathematical expression, to each theorem is obtained by turning around all arrows in that theorem [MacLane (1998)].

There are few other phenomena with such a universal general validity as chirality, probably none at all. Not even time with its direction from the past into the future is so generally valid. The photons for example, so important for physicists, are in a certain sense timeless, but chiral. This is the case also for the instantaneous non-local interaction between entangled particles.

Mathematically, chirality plays a fundamental role in category theory. Here chirality is represented by arrows ($\Rightarrow$) which are meant to symbolize a change, a morphism: each morphism has an origin and a goal. Mathematical morphism is therefore a good model for representing physical perceptions as a causal information flow from an object to a subject. A mathematical structure changes, whilst something else remains unchanged. How the change is interpreted mathematically or physically, as a time, location or structure of an object or as a relation, is left open by category theory. In category theory, each morphism can also apply in the reverse direction, even if it is not an isomorphism.

In set theory order as the duality of orientation is best expressed by the inequality of relations $(a, b) \neq (b, a)$ [Russell (1903)].

2.8 To Measure is to Count

The physicist must register his perceptions, that is, the information flowing to him from the outside, in such a way that it is unequivocally communicable. Unequivocally means that the information, as far as possible, is to be stored in the form of numbers, because there are no misunderstandings about numbers, at least as long as it concerns finite, rational numbers. This compels the physicist to measure. The measurement associates physical characteristics of objects with numbers which say something about the structure of these objects and/or about the perception thereof. This information is unequivocal only, if the relation between the chiral orientation of the numbers ($\geq$), the object and the perception thereof is clearly defined by a convention.

The measuring process is thus always a counting process. For this purpose, the physicist must know three things, namely which elements of an object he wants to consider, which relations between these he wants to measure and in which units he counts. During a speed check of motor vehicle traffic, the radar latches onto the rear of the vehicle and one assumes that the whole car moves at the same speed as the rear, although the individual atoms make completely different movements. The radar measures only the speed; the temperature relations and other parameters are of no interest. One measures in kilometers per hour. All physical measurements can be attributed to three kinds of counting: units of length, periodic events (for example of pulse beats) or non-periodic events (for example radioactive decays) are counted. There are no other measurements.

Since each measurement is based on an interaction between the object and the measuring apparatus, and because each information transfer from one object to another changes both of them, the physicist only knows the value of the object measured before the measurement. The physicist functions in this respect like a measuring apparatus with a consciousness which stores the received information as a document of the past. About the present, he has no precise knowledge.

During the counting of periodic events, the physicist compares the number of events with the number of pulse beats or the frequency of his internal clock. In so doing, he assumes as self-evident that the cycle duration between two subsequent equivalent events is always the same at the same place and in the same state of motion, although he cannot actually know anything at all about this duration. This is called the "chronometric convention".

Also the counting of units of length can be attributed to a counting of events, for example, the registered number of markings on a yardstick. Every kind of length unit can be derived from a time unit. However, strictly speaking the physicist does not know anything at all about the real distance between the individual markings. Here also one proceeds from a "geometrical convention", which in most physical theories reads: The length of a distance is independent of the location. That does not mean, however, that it is also independent of the state of motion of the distance. The chronometric and the geometrical conventions are necessary, but actually arbitrary agreements. Hence it follows that the physicist can never really measure and know length of time and distance [Grünbaum (1973), 3–65]. All he knows are counted numbers of events. That leads to the question as to whether length of time and distance are concepts suited to physics, or whether physics is formulated better and simpler by a theory about counted events. For this, however, one would need a practicable mathematical definition of the term event.

For counting, numbers and their axioms play an important role. For chirality theory, only finite rational numbers are required because all measurements are described by numbers of counted events. Thus, only few axioms are required, namely the simple Peano axioms. If surds, imaginary or infinite numbers were introduced, e.g. to describe waves or information, this would be part of the mathematical theory and not an aspect of real nature.

For the mathematical description of measurements a metric is required. Since infinity and continuum should be excluded in physical models the metric must define distances as rational numbers and it must be defined what entity is counted by these numbers.

2.9 Is the Probabilistic Outcome of Measurements Real?

Probabilistic outcomes of experiments happen for four reasons:

1. Depending on the definition of reality and existence something might exist but not be real (see definitions in Section 1). In chirality theory this is the case for single points whose only intrinsic property is their existence although they are not observable. This means that the presumed observer with his mathematical model describes an existing, finite, natural world whose structure is in principle not perceivable. Einstein proceeded this way: "God does not play dice". The real observer with his mathematical model then describes only his observations of real nature and makes corresponding probabilistic propositions. Bohr proceeded that way.
2. Every perceived object is changed by every perception. Therefore it is impossible to know the precise present state of any object.
3. In chirality theory there is one and only one universe. Every point is related to all the other points of the universe. A separation of the universe into systems or objects is always an artificial one and falsifies the outcome of the measurements. Nonetheless such separations are inevitable because the universe is too big and too complex to be observed as a whole and to be precisely described mathematically.
4. Measurements never are perfect.

All this is in harmony with the philosophy of quantum theory. The first two causes of probabilistic measurements are indeed related to quantum theory where the states have no definitive values for the measurement outcome. There is in fact zero-probability in nature itself, but not in the observation of nature. This is another example for the importance of strict distinction between nature, observation of nature and the mathematical model for the description of the observation.

2.10 Summary of the Metaphysical Conditions for Physical Theories

As a summary, the metaphysical conditions are to be enumerated which should comply with a physical theory, so that it is closer to reality than the conventional theories. In order to take seriously the physicist as a subject who takes into his consciousness empirical perceptions from the outside, the ego or "presumed observer" must play a central role in the theory. The theory should be simple, immediately evident and logical for the physicist. The things, which the theory describes, must be existing. For this, they must be causally linked. Perception is an ordered flow of information from the object to the subject, by which both of them change. Chirality, the duality of all Be-ing and the fundamental cause of every order, is an important basis for every physical theory. Infinity and continuum must not be mathematical tools of the theory. Quantities of information are the subject matter of the theory which makes statements about how such quantities change. Whilst something changes thereby, something else remains unchanged. The structures of object and subject must be such that the information is representable on it and that it can flow from the object to the subject. For this, the perceiving subject and the perceived object must be indeed connected with one another, but nevertheless clearly distinguishable. The unit of measure of the information is the binary choice, i.e. the answer to a yes/no question. These units are finite and computable as events. The mathematical model of the theory might be deterministic, but it must allow probabilistic outcomes of real measurements. Most of these basic conditions are only partially met by the existing physical theories.

Since ancient times, natural scientists have described the world with the help of the four entities of time, space, substance (= mass, energy, information) and interaction. Since Aristotle there have been repeated attempts to reduce the number of entities, the most recent being loop quantum gravity and string theory. But they have not led to a theory of everything and they cannot make general relativity compatible with quantum theory. They might work but they cannot metaphysically explain why they work. Nevertheless, analogies between the four entities seem to exist. Observed space and time are both chiral for the observer and they never appear independently of one another. The curvature of space affects matter and vice versa. Are space and matter really two different things? The photon with its particle/wave dualism is half particle, half interaction. Are space, time, matter and interaction really four different entities, or rather, are they four different aspects of the same entity? And what, then, is this entity? Can the mathematically appropriately defined event take over this role as new, only entity? Seen physically, there would then be nothing real other than events. Only they are perceivable. Only they exist. Yes, the act of the perception would be an event. To measure is to count events. The events would be the atoms of Be-ing. Is it conceivable that such a new physics, which is based on the event, leads to a theory of everything, with which the theories of general relativity, quantum theory and the four well-known interactions can be uniformly described? The reduction of the four usual entities space, time, substance and interaction to only one single entity, the event, is the fourth paradigm shift of chirality theory.

The search for a new physical theory should be methodically approached as follows. It begins with the formulation of a new mathematical language, logic and axiomatic, the only initial condition of which is that it should be as simple as possible and that the axioms do not contradict the mentioned basic metaphysical conditions, which are a precondition for every perception. Therefore, the axiom of infinity will be dispensed with and the axiom $A \equiv A$ will be replaced by a new axiom of chirality.

These first steps thus just take into consideration mathematical simplicity and basic metaphysical conditions. Very soon it will become evident that the mathematics thus developed will surprisingly also be interpretable as a natural law in the physical sense. Skillfully applied to physical perceptions, the new mathematics is apparently suitable for the description and unification of the theory of relativity with quantum theory.

3 The Model of Chirality Theory

3.1 One Point

In Section 3 a mathematical topological model is described without any physical interpretation of the model. Such a model is not "true", however it is considered useful for finding a mathematical definition of the term event which is practicable for physics and from which the terms space and time can be derived. It is beyond the scope of this article to set out a complete axiomatic. Such systems can be found e.g. with Stewart Shapiro [Shapiro (1997)]. Rather, we should begin with the simplest thing which mathematics has to offer as an ontological basic object: the point. According to Euclid, a point is a thing without parts. Alfred Whitehead's modern definition reads "A geometrical element is called a 'point', when there is no geometrical element in it" [Whitehead (1936), 456]. For the moment, it concerns neither a point in space nor one in time, nor even an observable object. There is no space to provide a background for the point. Indeed, there is not even an observer to assert that the point exists. But it must exist since existence is its only intrinsic property. It seems to be philosophically impossible to design a model of physical perceptions without anything that is not real but still must exist. The most simple, ontological, existing object that fulfills this condition is the point. Probably all other physical theories work with basic objects of much bigger complexity than chirality theory.

Single points are not perceivable because they do not contain any information. But relations between many points might be mathematically described in a way that a chiral structure arises which serves as the basis of order and information. At first sight it seems that a finite set consisting of n points should be introduced and examined what possible relations between all these points can be defined axiomatically. But it soon becomes evident that a set of n points is too complicated for defining a chiral structure in a simple way. Therefore, I will now add one point after another to the single lonely point and investigate what can be said at all about the model of the growing number of points.

The n-point system will be discussed later in Section 4.9. A mathematician probably would prefer such an n-point space or even a ∞-point space which would better correspond to our universe. But compared with the 4-point space the n-point space is much more complicated. It allows so many relations that the model would have to be simplified for calculations being made possible and for describing physical observations, e.g. by artificial perturbation and renormalization tools which have nothing to do with nature. This would contradict the metaphysical requirement of simplicity and compatibility with nature.

3.2 Two Points

The set of n points is a space. Two points can coincide or not. If they coincide, then the distance between them is zero or infinitely short. Since infinity is not perceptible and thus must be ruled out on metaphysical grounds, the two may not and cannot coincide. This must be secured by an *axiom of individuality*: Two points always can be distinguished by a presumed observer.

The relation between the two points I call *distance*. Within a space the points are no more independent of each other. There is a distance between them. The distance "not zero" between the two points can be assigned a number, for example, the number 1. Thus, there is neither a longer nor shorter distances between the two points, but just the distance 1. An *axiom of reflexivity* must ensure that the distance is independent of the direction in which it is measured. In a 2-point space neither chirality nor order can be defined.

3.3 Three Points

In a space consisting of three points, there are, thanks to the third point, possibilities for additional relations between the points; namely, the third point can lie between the other two or not. If no point lies between the other two, the three points form a triangle, whose sides all have the length 1. If the three corners have names, e.g. x, y and z, then they can be arranged clockwise or counter-clockwise in the space defined by the three points. Since points do not have names in nature, such a distinction is physically impossible, however. That said, there is another state, which is clearly distinct from a triangle, namely the case where a point y lies between x and z. For such a state an appropriate axiom of between must be formulated, something missing in Euclidean geometry as already Carl Friedrich Gauss had criticized. The axiom allows to describe locations in finite, n-dimensional topological spaces without a continuum, i.e. without requiring a mathematical "way" between the points. The locations on the surface, which is defined by the 3-point triangle, can all be characterized by the relation "between": The sides of the triangle consist of the set of all points which lie between the corners. The inside of the triangle consists of the points which lie between the 3 sides. The extension of the side beyond a corner consists of the set of all points with the characteristic that the corner lies between them and the side concerned. The points over a corner lie between the two sides extended beyond this corner. The points over a side lie between this side and the extensions of the 2 other sides. A point is "further on" or "outside", if it is neither one of the two points of the distance nor between them. The "between" will be used to define the term event.

The between-point y divides the distance xz into two parts, each of which has the distance 1. Thus, distance $xz = zx = 1 + 1 = 2$, and it is thereby clearly distinguishable from distance 1 in the triangle. In this way, all three conditions for a metric are fulfilled: the distance between x and x is zero and therefore quite invalid in physics, the reflexivity axiom is applicable and the distance xz is $\leq xy + yz$.

Now I permit the three points to move freely and to adopt different states in turn, both singly and as point sets. The individual points have four different states or possibilities of movement relative to the two other points. They are each characterized by an arrow symbol:

1. The point can move away from the two other points: (↦) = "move away".
2. The point can move towards the two other points: (⇤) = "move towards".
3. One point lies between the two other points: (↔) = "between".
4. The point can turn back: (↩) = "turn back".

(↔) designates the transition from (⇤) to (↦), whereas (↩) indicates the transition from (↦) to (⇤) So the individual point moves in a defined sequence through the four states: it moves towards the other two points, passes between them, moves away from them and then turns back. What causes it to turn back? It is important to always keep in mind that the points do not move in absolute space, but only relative to the other two points, which form the point space for the third point. This distinguishes chirality theory from most conventional theories of space which require a background space as part of the model. It is not points moving in space, but rather – since the points themselves form the space – space moving through itself. Each time a point moves through between the others, the space turns itself inside out like an inverted glove. If a point is situated at the tip of the glove's index finger and it moves away from the glove in the direction the finger is pointing, then as soon as the glove is turned inside out, this point will abruptly turn back to the glove and fly back into it. Exterior and interior have been transposed by the inversion of the glove. From the point's perspective, the direction of motion has not really changed; through the inversion of the space, the direction of motion, however, has reversed relative to the other two points, which form the surrounding space for it. The cause of the inversion of the space is always the point in the (↔)-state, i.e. the one which migrates through the other two.

The inversion of a point space I define as an event. The physical interpretation of the model will be given in Section 4. Events will be interpreted as the basic ontological real objects in physics. Should the physicist find these ideas plausible, then he can formulate the following *axiom of chirality*:

(⇤) ⇒ (↔) ⇒ (↦) ⇒ (↩) ⇒ (⇤) ⇒ and so on.

The axiom of chirality corresponds neither in the form nor the content to the usual mathematical or physical axioms. Since it portrays a process and consists solely of arrows, it belongs most appropriately to category theory, which can thereby probably play an important role in theoretical physics [Dümont (1999)]. The four states of motion of the individual points are thus ordered and have a direction.

The symbol (⇒) ("if ..., then ...") denotes in this context the phase change from one (motion)-state to the next.

The notions of "away", "towards" and "turn back" must be further defined. The expressions are illustrative for the picture of a glove that is inverted. But mathematically it is appropriate to define the arrows in a more abstract way: (⇤) is the state which leads to an event; (↦) does not result in an event; (↔) and (↩) mean that in these states, an event occurs whereby (↔) is the cause and (↩) is the consequence or action of it. This view has the advantage that the term event can be defined without requiring the term distance: "to move towards the other points" no longer means that there is a distance between the points which becomes shorter, but simply that the motion leads to an event.

Distances only ever change as a result of events, thereby becoming countable.

The axiom of chirality is intended to make propositions about many points. In the definitions of the five different axiomatic arrows there were only 3 points. To become applicable to an n-point space the axiom of chirality must be somehow mathematically modified (see Section 4.9.5).

3.4 The 4-Point Space

For the 4-point space the axiom of chirality describes the relations between 4 points and the changes of these relations from phase to phase, i.e. from event to event. The points themselves have no other property or meaning than the fact that they exist. When not all of the 4 points lie on the same plane, then they form the corners of a tetrahedron, i.e. a three-dimensional space composed of 4 points. Here also, the points can adopt the 4 different states of motion (⇤), (↔), (↦) or (↩). (The notion of dimensionality will be discussed in Section 3.5.) If one of the 4 points is in the between state relative to the other 3 points, the tetrahedron is inversed like a glove in 3-dimensional space. If each of the 4 points is assigned another of the 4 possibilities of movement relative to the three other points, then there is after each phase change one point in the (↔)-state and the process of space inversions following one another never stops. The chain of events cannot break down, and after every event the result is once again a tetrahedron with the 4 corners in the 4 different states. Only a disruption from outside would be able to stop this process. An outside does not exist in the 4-point space, however.

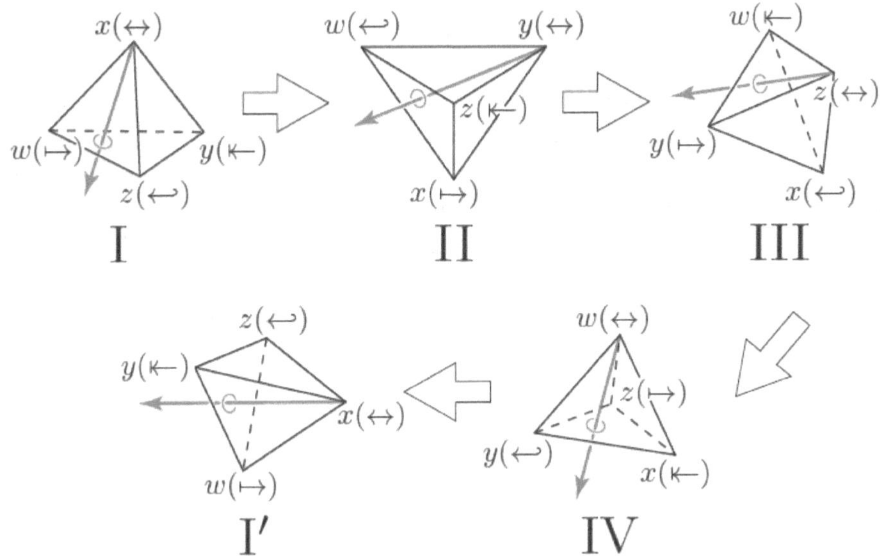

Fig. 1 4-point space. *Following phases I to IV, the 4-point space returns to the original state* **I'**, *although in another position from the perspective of the external, presumed observer. If one disregards the designation x, y, z, and w of the four corners, then the state of the 4-point space is the same following each phase change, only its position has changed, whereas the chiral orientation has remained the same: viewed from the* (↔)*-point, the triangle* (↦), (↩), (⇤)*always rotates in an anti-clockwise direction.*

In Figure 1, such an unrestricted sequence of periodic events is illustrated. Both the tetrahedron xyzw as well as its process of movement are *chiral*. That is to say, image and mirror image of the tetrahedron and of its change process cannot be brought to coincidence with each other. The chiral orientation remains the same from phase to phase. The tetrahedron thus retains its *identity* during the eternal process, i.e. all its intrinsic characteristics remain unchanged, although all the relative states of the single points have been changed by each event.

From the perspective of an external presumed observer, the motion of the 4-point space is *isotropic*, i.e. no direction is distinguished in his 3-dimensional space.

For the (↔)-point, there are 3 possibilities. It can pierce the surface defined by the other three points in three different locations: the surface in the triangle's interior, or above a side or above a corner (Figure 2).

These locations on the surface, which is defined by the 3-point triangle, can be characterized by the relation "between" as shown in Section 3.3.

As long as all 4 points are in the 4 different states there are consequently six versions of the 4-point space which can be differentiated by their intrinsic characteristics: each of the 3 possible states of the (↔)-point can occur in 2 chiral variations with different orientations.

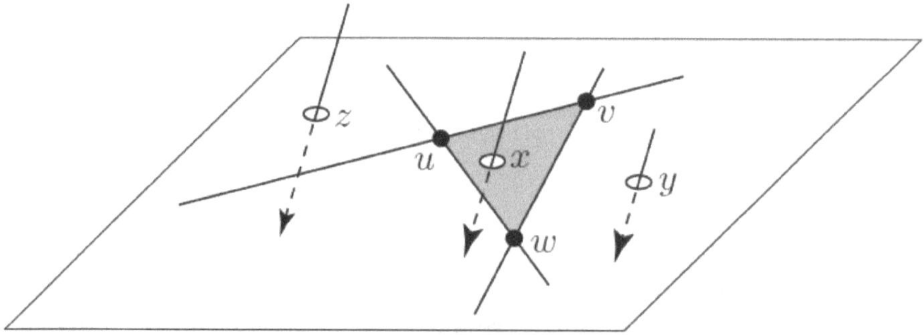

Fig.2 The three possibilities for the point "between". *The plain which is defined by the triangle (u, v, w) can be pierced in essentially three different locations by a fourth point: inside the triangle $x(\leftrightarrow)$, above a side $y(\leftrightarrow)$ or above a corner $z(\leftrightarrow)$.*

If the 4 points are not in the 4 different states relative to the other 3 points, the process of events cannot be eternal and the space becomes chaos.

3.5 Dimensionality

The dimension of a mathematical space or object is informally defined as the minimum number of coordinates needed to specify any point within it. For metric vector spaces this is the number of countable basis ($\mathbb{R}^n$). The problem of the meaning of dimensionality with the 4-point space of chirality theory is threefold:

1. Because there is no manifold as background space there is no countable basis ($\mathbb{R}^n$).
2. The 4-point space inverts with each event.
3. Since there are only events to be counted it is not evident why the dimensionality can be other than 1.

It follows that in chirality theory other criteria must be found for determining the dimensionality of a space. This is not unusual in mathematics where various definitions for the notion of dimensionality are used. This situation resulted in rational, irrational, imaginary or negative numbers for the dimensionality in specific theories.

Since the 4-point space consists of 4 points in 4 different physical states relative to the states of the other 3 points the 4 points can be characterized or illustrated by a tetrahedron. Each of the 4 points requires a dimension of its own to distinguish it from the other points. The single first point is 0-dimensional; together with the second point a 1-dimensional space with a distance 1 between the points is formed; the third point gives rise to a 2nd dimension and with the fourth point the space becomes 3-dimensional. A set of n points in n different states relative to each other results in an $(n - 1)$-dimensional space. The 4 points in 4 different physical states are the 4 basic *existing* ontological objects. The basic *real* ontological object is the 4-point space as a whole. Only this is perceivable and one wants to know how many dimensions are required to describe this observation mathematically.

According to the axiom of chirality each point changes its state with each event whereas the space as a whole remains unchanged. How this change is accomplished, by an inversion or by a 3-dimensional rotation of the space, is left open and cannot be observed. By the inversions of the space the events produce a chiral rotation of the space that can be described by two polar coordinates φ and ψ as shown in Figure 3.

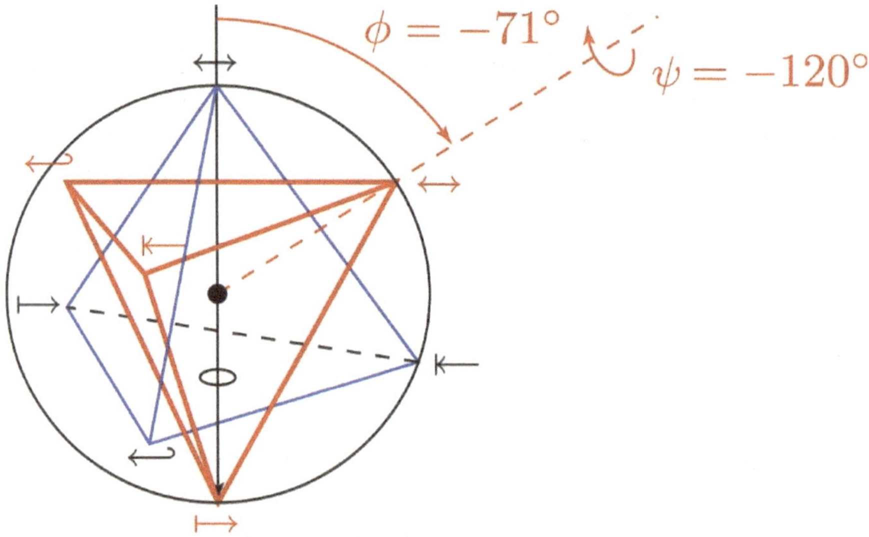

Fig. 3 Chiral rotation of the tetrahedron in 3-dimensional space.
Φ = clockwise rotation about the axis through the centre of gravity (parallel to the blue side of ↩ to ↦) of −71°. Ψ = clockwise rotation about the axis through the centre of gravity and ↔ of −120°. The inversion of the blue tetrahedron relative to the red one corresponds to a rotation through the angles Φ and Ψ.

To describe such a rotation a 3-dimensional space is required and the orientation or rotational direction of both angles must be defined. The events can be counted. Such a chiral process with four different point states is only possible in a 3-dimensional space. In a 2-dimensional space no tetrahedron could be constructed and in 4-dimensional space the process would not be chiral any more.

With the between-state as it is drawn in Figure 2 the drawing itself is 2-dimensional. But the drawn space is 3-dimensional because the between-point is not only at a location of that space, it also has a direction (arrow) perpendicular to the 2-dimensional space formed by the other 3 points. For that a third dimension is required.

It must always be made clear the dimensionality of what is considered. In this section it was the dimensionality of the model of a 4-point space obeying the axiom of chirality. If instead nature or the observation of nature is described there might be other dimensionalities (see Section 4.3).

4 Physical Interpretation

4.1 The Presumed Observer

A real observer is himself subject to the laws of nature. By means of his real measuring apparatuses he describes his empirical observations. He counts the perceived events. As shown in Section 2.1 a presumed observer who describes the system of both the observer and the observed events together is inevitable for metaphysical reasons. For the moment it is assumed that one and only one such presumed observer is present. He has an inner clock and can count, but the counting is possible without any interaction. The real observer is physical and real, whereas the presumed observer is as transcendental as his mathematical model.

The presumed observer can view the 4-point space from two different locations: either he perceives the tetrahedron from outside as a whole how it inverts or rotates chirally in leaps from event to event (Figure 3), or as an internal observer he accompanies one of the 4 points, and perceives how it repeatedly pierces the plane defined by the other 3 points and proceeds through a translation. Thus, in the 4-point space, the observer is located either outside or inside the black mini-hole (see Section 4.3).

In order to count events the presumed observer must have a transcendental structure that enables him to order information. To be able to do that he must be chiral. This ordering structure is most simply defined as an internal clock with a subjective time direction. But if the theoretical physicist prefers some other chiral structure as e.g. the transcendental ordered letters of a book he is free to do so. There is no need of an internal clock or a background time in chirality theory, at least not more than there is a time in every mathematics because the mathematician must have time for his thinking. For the description of a sequence or for a causality an order or an arrow is required. The metaphysical source of the arrow is the flow direction of information from object to subject. This flow must be ordered, i.e. chiral. This does not have to be an arrow of time. It might be any chiral duality.

The relation between the chiral order of the presumed observer, i.e. the arrow of time of his internal clock, and the order of the counting process must be defined by a convention. If the counted events are periodic, they are the measure of time (not time itself). If they are not periodic, they are the measure of space (see Section 4.4). The direction of the time arrow and the orientation in space are subjective. They are not an inherent aspect of nature. So the time arrow of a perceived antiparticle might be opposite to the time arrow of the observer. Mesons consist of a quark and an antiquark. This means that within the same meson time might flow in two opposite directions as postulated by Richard Feynman [Feynman (1985)]. So the presumed observer "lives" in his own time with his subjective time arrow. But to order his perceptions he uses a time model that allows both directions of the arrow. Time *per se* is not an inherent aspect of nature (see [Einstein (1979)]). The inherent aspect is chiral order.

4.2 A Neutrino Model? Definition of Elementary Particle

The mathematical model which has now been presented is not an end in itself. Rather, we are actually looking for models which on the one hand, come as close as possible to the basic metaphysical conditions, and on the other hand, are able to describe physical objects and perceptions thereof. The presented model resembles rather the models of atoms and molecules in chemistry than the description of particles as it is used in the *Grand Unified Theory* (GUT). The advantage of the model is that it is easier to understand and to be imagined or to build up more complicated particles such as quarks and nucleons out of the building stones. The disadvantage is that the model does not (yet) allow any calculations.

The simplest known particle today is the neutrino. As is the case for the presented model of the 4-point space the neutrino occurs in six different versions: as an electron-, muon- or tauon-neutrino with a left-handed spin or as one of the three corresponding antineutrinos with right-handed spin. Neutrinos and antineutrinos are distinguishable by their opposite orientation, as in the model of the 4-point space. So this model describes the intrinsic characteristics of the neutrino quite well. An explanation for their varying masses remains open. Possibly the bigger mass of the muon- and tauon-neutrinos is due to the longer radius of their (↔)-point during the space rotations. According to this model the neutrino as an elementary particle consists of 4 related points with a dynamic eternal process regulated by the axiom of chirality.

I define *elementary particles* as objects consisting of a definite number of related points which move periodically relative to each other according to the axiom of chirality.

4.3 The Dimensionality of Physical Space and Black Holes

The ideas of this section are speculative. The described possible extensions of chirality theory are discussed in my book "Metaphysics. Chirality as the Basic Principle of Physics" (2008) [Wehrli (2008), 159–271].

Since, according to the axiom of chirality, a point can only take up one of four different states of motion, only in 3-dimensional space is it possible to construct a chiral tetrahedron with four different corners in such a way that the internal chain of events never stops of its own accord (see Section 3.5). This is probably the underlying reason why physical space as we believe we perceive it is 3-dimensional. Three is not the dimensionality of nature itself but that of the perception of nature or of the *apriori* space of the "*Anschauung*" as it was postulated by Immanuel Kant (see Section 5.17). In 4-dimensional space the chiral tetrahedron of the 4-point space could not be distinguished from its mirror image and perception would therefore be impossible. Every empirical perception is probably based on precisely this possibility of distinction. Only 3-dimensional space permits perception, and without perception there is no physics.

4-point spaces might have been created by a spontaneous symmetry breaking of a system consisting of many chaotically moving points as it possibly existed shortly after the big bang. No point can leave the 4-point space in which the 4 points are in the four different states. Whenever the (↦)-point looks to break out, the space inverts and the point moves towards the other 3 again. This way there exists an attraction between the 4 points analogous to the gravitational attraction in physics, whereas repulsion is impossible. Points are attracted due to the inversion of 4-point spaces.

A 3-dimensional space from which no point can escape, because the space inverts too quickly, I define as a *black hole*. A black mini-hole is stable because of its high symmetry. As long as there are not more than 4 points there is no way for it to decay. But if there is a fifth point, that point could replace one of the other 4 points thereby producing another type of neutrino, e.g. a muon-neutrino out of an electron-neutrino. In contrast to the 4 points of the 4-point space a fifth point does not move in a defined manner. Single points might travel chaotically through our universe. When such a point gets in between the ordered points of the black mini-hole one of the 5 points will have to escape that mini-hole because the neutrino is stable only as a 4-point space. Such "oscillations" between different neutrino flavors have been experimentally observed especially in regions where there is dense matter as is the case in the interior of the earth [McDonald et al. (2003)]. It is not possible nor is it necessary to describe a mathematically precise mechanism of this oscillation.

Presumably the definition of the black hole could be extended to include spaces with more than 4 points. The neutrino model is the smallest possible black hole. An extension of the notion of the black hole might be the case where four black mini-holes together form during a second symmetry breaking a new black hole the same way as the neutrino model had been formed by 4 single points. For this possibility $4 \cdot 4 = 16$ points would be required. Such a system could be interpreted as an electron or positron. Until now no such electron structures have been experimentally observed. This is no surprise because it is impossible to observe in the universe or produce in the laboratory temperatures and densities as they must have existed during the big bang. The action of these new particles could be as well attractive as repulsive depending on the relative orientation of the black mini-holes.

Also other elementary particles such as the 72 different quarks and antiquarks, or mesons and nucleons might be formed in a similar way according to dynamical rules that all can be derived from the axiom of chirality. Such an extension of chirality theory would result in the standard model of the *Grand Unified Theory* (GUT).

In this section the dimensionality of nature itself is described. This means that also the dimensionality of the black holes must be included. Every symmetry breaking that leads to a new level of black holes also gives rise to 3 more space dimensions. This is because only 3 space dimensions are observable and no information can be transferred from the inside to the outside of a black hole. Therefore, there are in nature a 1-dimensional time +3 dimensions of space +3 concealed dimensions of black mini-holes +3 concealed dimensions of black mini-holes within the black mini-holes = 10 dimensions, the same as is postulated by some string theories. The 10 dimensions cannot be observed empirically but they must exist. In both chirality and string theory 6 of these dimensions are "concealed". In chirality theory, this is due to the black mini-holes from which no point and no information can escape.

Even the universe itself could be the interior of a black hole. The not perceivable, chaotically moving single points in such a universe then could be interpreted as the so called *dark matter*, which contributes about 27% to the total mass of the universe.

The extension of the notion of black holes is the fifth paradigm shift of chirality theory.

4.4 Space and Time

The external presumed observer counts the events, i.e. the periodic inversions or rotations, of the 4-point space. The event sequence and the counting thereof are one-dimensional and chiral, that is, they have a direction. The dimension in which the periodic events are counted I define as *time*. The number of periodic events is the duration of time, which is thus a dimensionless number. Periodic means in this context that the object, i.e. the 4-point space, is in precisely the same intrinsic state before and after the event. This definition of time uses the term "event", whereas the entity event has been explained without using the word "time". So the event is more fundamental to physics than time.

The time arrow of the observation is that of the internal clock of the external presumed observer. This means that the direction is subjective and relative. The observer cannot know anything about the direction of the time arrow within the observed neutrino or antineutrino. He only knows the number of counted periodic events and that number is the same independently of the time direction. Since after each event the neutrino looks exactly the same as before the event, there is no evolution of the observed object which could indicate a time direction. It might be that the time arrow of the neutrino is opposed to that of the antineutrino.

The internal presumed observer "experiences" the events as the piercing of the plane defined by the other 3 points. These events are a change of location via a translation. The point reaches "a location further on" by passing a (↔)-state. These changes of location are the countable quasi markings on his yardstick. He moves with his point from location to location through the space of the 3 other points, a space which from the perspective of the internal observer rotates. Thus, he counts only every fourth event in the 4-point space; there are 4 such possible internal observers, one for each point. The dimension in which the changes of location are counted I define as *physical space*. Since, from the external observer's perspective, the 4-point space rotates without a direction in space being especially distinguished, we get, viewed from the outside, the impression of an isotropic, 3-dimensional space. (Regarding the notion of dimensionality see Section 3.5). In this space, the number of location changes is a measure of the internal observer's distance covered. Since this measure is a number of counted events, distances involve dimensionless numbers. Seen from the outside the point stays all the time within the black mini-hole, i.e. at the same location. Observed by the internal presumed observer the point travels a long distance from event to event. Neither time nor space and distance are real. Real are only the events. Put succinctly: whether an event is perceived as a change in space or a change in time is relative and depends on the standpoint of the observer. What the observer outside of a black hole perceives as a length of time is for the observer inside the black hole a distance. For the object itself, there is no difference between time and space, there are only events.

A *clock* is a periodic event counter. It runs most quickly for an observer who perceives and counts all events as rotations. The *standard length*, however, measures or counts changes of location. Every event which it does not count as a periodic rotation is lost to the clock, because the same observer cannot count an event twice, partly as time and partly as a change of location. Consequently, there is a fastest possible time and a longest possible distance. The observer measures the fastest possible time when he counts all events as rotations or periodic space inversions and the object is at rest. He measures the longest distance when he perceives all events of the observed object as changes of location, i.e. when time stands still for the object. Such objects are, for example, the graviton or the photon. This state of affairs corresponds precisely to the conclusions of the special theory of relativity.

4.5 Velocity *c*

The ratio of the number of location changes to the number of periodic time events is called *velocity*. The fastest velocity is measured when all events are counted as location changes and this number is compared with the event count of the presumed observer's inner clock. The corresponding velocity is called *light speed* c. It is the natural and unambiguous measure for all velocities. The observer's inner clock can be calibrated such that the ratio c of both numbers becomes 1. In this case, every event which is perceived as a change of location corresponds to a time event of the observer's inner clock:

$$c = 1{:}1 = 1.$$

In 4-point space this means that the inner clock of the presumed observer always counts a time event of his inner clock whenever he is between the other 3 points. Consequently, the points inside the tetrahedron move at light speed. Since velocity is nothing more than a ratio of counted events, it is also a dimensionless number.

In this section the velocity is described and defined as that of a single point as it is observed by a presumed internal observer and not as that of a 4-point space as it is observed by a presumed external observer. A presumed external observer would not see any distances but only periodic events, i.e. time. But there is no presumed external observer. He would not see any velocity at all because for him the observed 4-point space is at rest (velocity = 0). The same observer never can be inside and outside of the 4-point space i.e. of the black hole. And if there were two such observers, communication between them would be impossible.

4.6 Frequency ν and Mass m_0

The ratio of the number of observed events (periodic events and location changes collectively!) to the event count of the presumed observer's inner clock I define as *frequency* ν. The minimum frequency during the observation of an object is measured when no change of location is observed, i.e., when an object is at rest. This minimum ratio I define as *rest mass* m_0. The frequency or rest mass of such a stationary particle is the result of the particle's internal events. It is thus independent of its state of motion. Mass and frequency are also dimensionless numbers. The external presumed observer's inner clock can be calibrated such that it always counts an internal time event whenever one of the neutrino model's 4 points is located between the 3 others. This calibration results in the lightest neutrino's rest mass becoming 1.

4.7 Spin, Fermion and Boson

The tetrahedron itself, as well as its rotational direction, is chiral during every event, that is, it cannot be brought to coincidence with its mirror image. *This chiral 3-dimensional rotation I define as spin.* The term 3-dimensional rotation is intended to express that the motion does not require a fixed axis or direction of motion. So the spin is not an angular momentum, but rather the abstract consequence of a chiral particle having precisely two possible orientations in 3-dimensional space. A neutrino, for example, has no axis and as long as it is at rest, also has no direction of motion. Nevertheless, it has a spin. Only with electrically charged particles and with the large systems of classical mechanics does spin manifest itself as an angular momentum with an axis [Feynman (1966 b)]. Spin states arise from a process of events. The orientation of the spin in a 4-point space is conserved because the eternal process only could be interrupted by an external action.

The rotation of the space which is observed by the external presumed observer comes about as the two points (⇤) and (↦), with a distance 1, fly right past each other at light speed whilst the (↔)- and (↩)-points remain at rest in this phase. It is thus

$$c \cdot c = c^2 = (1{:}\,1) \cdot (1{:}\,1) = 1$$

This description of the 3-dimensional rotation is intuitive. For a more precise description of the 3-dimensional topological changes, one would require an axiomatic system which is yet to be formulated. The equation can also be interpreted as the change in the two angles φ and ψ per event. The complete 3-dimensional rotation is the product of the two changes of the angles. The chiral orientation of the rotation can be portrayed by allocating the rotation an algebraic sign $\pm$. By means of a chiral, 3-dimensional object, for example a right hand, it must then be defined which orientation is to be positive and which is to be negative.

In order to fulfil the basic metaphysical condition, that the flow of information from the object to a real (instead of presumed) subject changes both of them, the orientation of the spin must change during the perception, that is, it must reverse. So the real subject can never perceive the 3-dimensional rotation itself, but only its change from before to after the measurement.

Mathematically, this can be expressed most simply such that the perceived, minimum change to the rotation, namely quantity 1, is the difference between $-\frac{1}{2}$ and $+\frac{1}{2}$, whereby the quantity $\frac{1}{2}$, which is not perceivable and real *per se*, is defined as the angular momentum of the 4-point space and the particle with the spin $\pm\frac{1}{2}$ is defined as a *fermion*. This way, the spin orientation becomes a binary choice, plus or minus. Fermions are only indirectly observable but real.

The four internal presumed observers together count the same number of events as the external presumed observer alone. So the individual internal presumed observer counts only $\frac{1}{4}$ as many, namely those when he is in the (↔)-state. The rotation of the 4-point space as a whole must be, seen from the perspective of the internal presumed observer, just as big as it is from that of the external observer. However, since within the 4-point space he only perceives every fourth event as such, from his perspective the rotational motion from event to event is 4 times greater than it is for the external observer. It is

$$4 \cdot \frac{1}{2} = 2.$$

This spin, like that of the fermion, can be positive or negative. So it is ± 2. Particles with integral spin I define as *bosons*. Bosons with their integer spin are directly observable, countable and real. The boson with spin ± 2 described is probably the graviton. Thus, the same 4-point space has, from the perspective of the external observer, a spin of $\frac{1}{2}$ and is therefore a fermion, whereas the spin from the internal observer's perspective is 2 and the object is a boson. The internal observer measures a 4 times greater rotation per time event of his inner clock than the external observer. Put succinctly, whether this particle is perceived as a fermion or as a boson is relative and depends on whether the observer is inside or outside the space that is the black mini-hole.

With the chirality theory, the inversion of the 4-point space during an event is accomplished by an inversion of the model in which space and observed object are the same. The situation is similar to that of general relativity where space and mass also are the same entity. The inversion of space is observed by the presumed observer and is therefore not real. A real observer could observe an object only, if the spin of the object is changed. In fact he would not observe the object as such but the inversion of its spin from $-\frac{1}{2}$ to $+\frac{1}{2}$. After the observation by a real observer the orientation of the spin is changed.

4.8 Planck's Constant *h* and Energy *E*

Central to the description of the spin was the equation

$$c \cdot c = c^2 = (1{:}\,1) \cdot (1{:}\,1) = 1 = h.$$

This constant h I define as *Planck's constant*. Since h is nothing other than a ratio of counted events, *h* is also a dimensionless number, namely 1. h is the smallest possible angular momentum an observer can perceive, since at least one event must take place for a rotation to come about. There are no half events. With c and h, one and the same physical constant is involved, which is perceived by the observer outside a black hole as the minimum angular momentum; by the internal observer on the other hand as the maximum velocity of the point he accompanies. Since in our universe c can be observed and measured the question arises, whether our universe should be interpreted as the interior of a black hole. In that case the surface of the horizon of the black hole would – seen from the interior of the universe – correspond to the moment of the big bang.

If the mass m_0 of the 4-point space is synonymous with the frequency ν and $c^2 = h$, then

$$m_0 \cdot c^2 = h \cdot \nu = E.$$

This equation unifies the special theory of relativity with quantum theory. The quantity E I define as *energy*. This is also a dimensionless, rational number.

4.9 The *n*-Point Universe: Particles, Distance, Action and Information Transfer

The model of chirality theory described up to now primarily considers the simple 4-point space. This model illustrates the philosophical idea of the theory quite well but of course it cannot describe the universe as a whole. As explained in Section 3.1 the description of an n-point space is mathematically difficult, so that artificial simplifications have to be made that contradict nature. Nevertheless, in this section some rules regarding the mathematics and the interpretation of an n-point space are sought. Today only a qualitative description of terms such as distance, action and information transfer in a space with more than 4 points can be presented. For a mathematician such "poetic" models might be unsatisfactory but I hope that some scientists will be motivated to think in the same direction and develop useful axioms and a complete mathematical theory for the description of physical observation. Even in the present form chirality theory can already metaphysically explain many physical phenomena despite the mathematics of the theory being rudimentary. This is in contrast to other modern theories with their beautiful mathematics which are difficult or impossible to be applied to physics and which cannot explain any physical phenomena at all.

4.9.1 Universe

The universe consists of an unlimited number of points. For forming an ontological existing object these must all be related to one another as well as to the physicist as the observer. An infinite number of points would contradict the metaphysical condition of being perceivable. All points are related to each other by their relative position and changes of position.

Every separation of points, particles, objects and subject is an artificial one and might falsify the description of a perception. Nevertheless such separations are inevitable in physics because the universe as a whole is much too complex to be precisely described. Therefore, mathematical technics must be developed for simplifications which do not unduly falsify the result. Such technics are e.g. used in renormalization and perturbation theory. These technics are never aspects of nature but only mathematical tools to simplify the description of real nature. It is important for the physicist to remain conscious of the ontology of his observations and not to believe that the simplified description of nature is truth. It cannot be a task of this article to discuss such mathematical tools and they have still to be further developed.

4.9.2 Space, Location, Distance, Event

Physical space is formed by any subset of $n \geq 4$ points. In a system of n points there are $\binom{n}{4}$ 4-point spaces. They can overlap. Sets of only 3 or less points cannot describe a real system obeying the axiom of chirality and therefore cannot be a model of reality. Space is not an aspect of nature, it is only one of the model. In that model the 4-point space is 3-dimensional. As long as we describe our physical measurements of space and time by the chirality theory model we describe 3-dimensional objects. This does not mean that the universe as a whole is 3-dimensional because there might be objects within objects called black holes. With every new level of such black holes 3 more dimensions arise (see Section 4.3).

Locations are defined by an *axiom of between*. An event occurs when a point gets between any 3 other points thereby changing its location. Thereby the 4-point space is inverted. Events are countable by natural numbers. *Distance* is the relation between any 2 points of the universe (see Section 4.9.5).

4.9.3 Measurement. Periodic and Not-Periodic Events

Events might occur periodically or not. A physical measurement consists either of the counting of periodic temporal events by a clock or of not-periodical counting of events by marks on a measure stick. The relation between the number of counted events and the physical unit of the distance or time has to be based on a convention.

4.9.4 Single Points and Particles

Single points might travel chaotically through our universe. Possibly they are the source of the mysterious dark matter. No single point can leave the universe. If the universe is interpreted as the inside of a black hole the point cannot spontaneously leave this black hole. It remains related to all the other points of the universe. Regarding dark matter and the possible role of single points for the flavor oscillation of neutrinos see Section 4.3.

If a set of points forms a particle at rest, i.e. one with exclusively internal periodic events, then the internal dynamics and properties of the particle can be described as explained in Sections 4.2–4.8. All dynamical laws must be derived from the axiom of chirality. The axiom postulates that the inversion of a space consisting of 4 points in the 4 different possible states relative to one another prevents the point in the (↦)-state from "moving away".

4.9.5 Distance. Modification of the Axiom of Chirality

All points of the universe are related to one another by a distance.

In an n-point universe there are many 4-point spaces. Every point belongs to many of them at the same time and the spaces may overlap. To account for this complex situation the axiom of chirality has to be appropriately modified. Such an extension of the axiom of chirality still has to be sought.

According to the axiom of chirality for the 4-point space the distance is the number of counted events during a translation. This should also apply to the n-point space. But there the weight of the inversions of the numerous different

4-point spaces depends on the distances between the points within the involved 4-point spaces and also these distances are countable. The larger they are the smaller the weight or action of the involved event.

With this model two types of distances have to be considered, namely the counted events of the presumed observer who travels by the shortest way along the measured distance and the distances between this observer and all the other points of the universe. Since in chirality theory all distances are counted numbers of events, the distance between two observed objects S_m and S_n must be a counted ratio $\mathbb{Q}_{m,n}$ of the 2 types of involved non-periodic events. This description is very vague and intuitive and the algorithm to calculate the distance is not yet known but in the end every distance must be a finite rational number of counted events.

This way the measured distance depends on the density of the observed system analogous to general relativity theory in which at the Schwarzschild radius of a black hole all distances become zero. But – in contrast to general relativity theory – in chirality theory there are space inversions and no space continuum.

4.9.6 Non-Local Action and Information Transfer

According to the axiom of chirality every inversion of a 4-point space with the 4 points in the 4 possible different states relative to each other exerts an attractive action onto all other points of the universe. The action is the result of a change of the configuration of all involved 4-point spaces. Such changes of configuration always happen instantaneously with every event. No point must thereby travel from S_m to S_n. Therefore, all actions are instantaneous. No time is required. But since the system S_n has no information about the changes in the system S_m it does not "feel" the action (Figure 4).

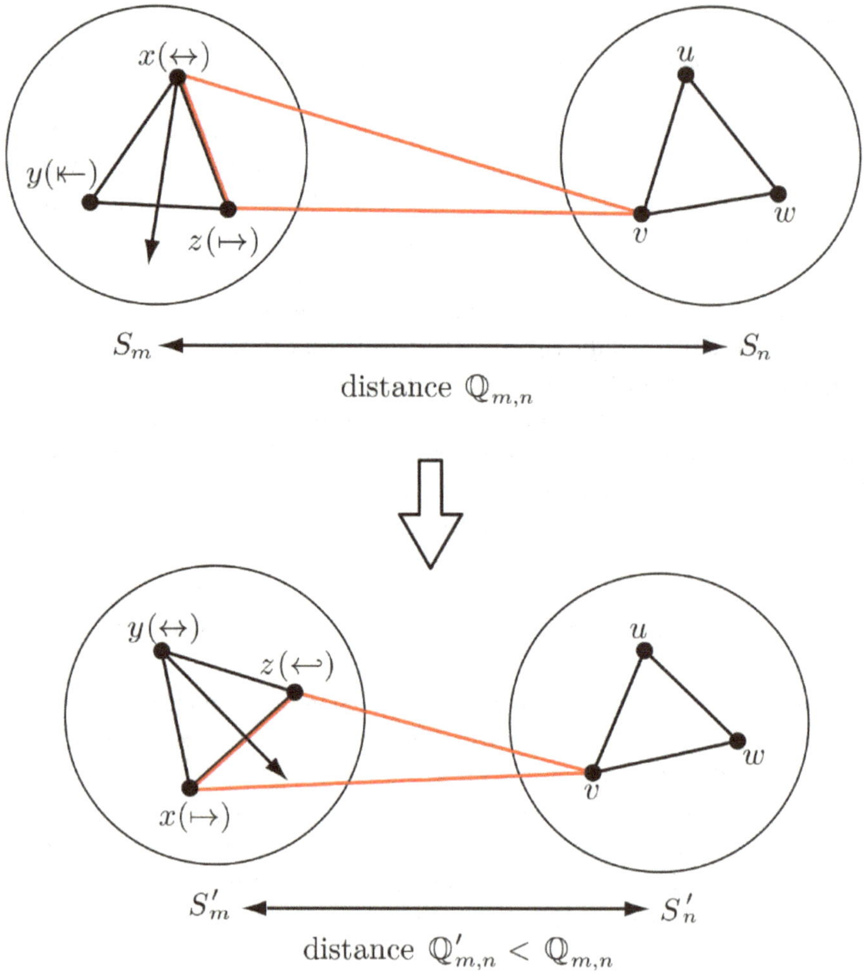

Fig. 4 Instantaneous, non-local (inter)action between S_m and S_n.
The orientation of space (x, v, z) inverts during the event to (x, z, v), but all points stay within the original systems S_m or S_n. Before $v_{(\mapsto)}$ in the red space is transformed to $v_{(\hookleftarrow)}$ and later to $v_{|\leftarrow}$ and $v_{(\leftrightarrow)}$ many phase changes ($\Rightarrow$) are required depending on the distance between the systems S_m and S_n. The third dimension of space and the weave of all the overlapping 4-point spaces between S_m and S_n are not shown.

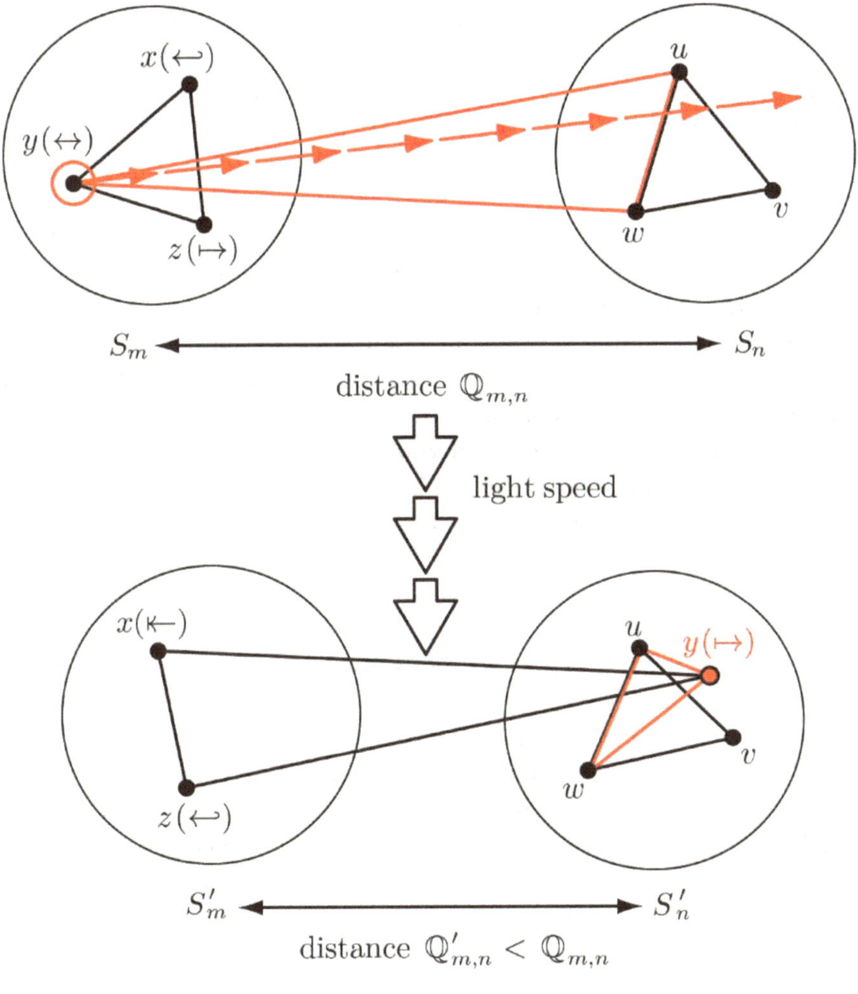

Fig. 5 Information transfer from S_m to S_n with light speed.

Point y (or the wave it has produced respectively) travels with light speed from S_m to S_n whereby both S_m and S_n are changed. The orientation of space (y, u, w) inverts to (y, w, u). The third dimension of space and the weave of all the overlapping 4-point spaces between S_m and S_n are not shown.

For an action to be perceived the total system ($S_m + S_n$) must be considered and at least one point must travel from system S_m to system S_n. Only this allows a change of the internal configuration of the system S_n that it can "feel". Since according the axiom of chirality all points travel with light speed the information cannot be transferred faster than light (Figure 5). This does not mean that the same point travels all the way from S_m to S_n. The information transfer happens through a net or weave of partially overlapping space inversions of all the 4-point spaces between the two systems. From this it follows that actions always are non-local but information transfer between separate systems is always local. It must be kept in mind that every such separation of systems in the universe is an artificial one. This means that the distinction between local and non-local is a subjective one depending on the extent of the artificial separation of the observed systems. These interpretations of action and information, locality and non-locality are the sixth paradigm shift of chirality theory. The questions regarding the velocity of action have been extensively discussed in the Stanford Encyclopedia of Philosophy [Stanford Encyclopedia of Philosophy (2008)].

4.9.7 The Waves of Quantum Theory

The information or energy transfer through a net of partially overlapping 4-point space inversions might result in a wavelike transfer at light speed as is postulated by quantum theory. An inversion of one of these innumerable 4-point spaces may exert an action in many other overlapping 4-point spaces which share one of their points with the previous 4-point space. According to the axiom of chirality this action produces new inversions of these other 4-point spaces depending on the relative configuration of the spaces and so on. The description of all these numerous inversions is too complicated to be drawn in a figure but the picture probably would resemble that of a wave as in quantum mechanics. But in contrast to quantum mechanics there is no background space for the wave and no continuum with a local neighborhood of the single points.

5 Comparison of Chirality Theory with Other Theories

5.1 Physical Theories

During the last 3000 years, humanity suggested many theories to describe the world by natural laws. Whole libraries have been written. It is impossible to present an extensive comparison of all these theories and to name all their discoverers. Nevertheless, it is possible to compare the philosophical and mathematical methods and notions which were used. Such comparisons have been published among many others by Aristotle (384–322 b.c.) [Aristotle (1984)], F. Capra [Capra (1991)], P. Davies [Davies (1989)], A. Einstein and L. Infeld [Einstein and Infeld (1950)], R. Feynman [Feynman (1965)], D.R. Finkelstein [Finkelstein (1996)], B. Greene [Greene (1999)], A. Grünbaum [Grünbaum (1973)], T. Kuhn [Kuhn (1967)], R. Penrose [Penrose (2004)], C. Rovelli [Rovelli (2010)], L. Smolin [Smolin (2001)], G. 't Hooft ['t Hooft (2015)], C.F. von Weizsäcker [Weizsäcker (1985)], A.N. Whitehead [Whitehead (1936)] and by the author [Wehrli (2008)]. All of them have influenced my chirality theory. Although there are fundamental differences between chirality theory and all the other theories there might be useful analogies which could be used to further develop other modern theories. To find a metaphysical strategy for the development of a physical theory is a difficult task. How this can be done starting from a formal language has been described by A. Döring and C.J. Isham [Döring and Isham (2007)]. They also have inspired my work although I could not follow all their ideas.

Of course, the strategy of chirality theory has not yet been completed and there are many open questions as is also the case for all the other theories. But a few fundamental principles that differ between chirality theory and many other physical theories shall be specified in this section.

5.2 Subject

The subject who observes and describes the universe might be

- a transcendental consciousness
- a subset of the universe which is also governed by the laws of nature
- a single ego
- a subject which can communicate with other subjects thereby giving rise to objectivity.

In chirality theory, one and only one subject with consciousness, the ego, exists. Since another subject never can unequivocally be differentiated from an object, every observation is actually subjective, even when the subject is not conscious of that. Every physical observation is a single and subjective one because an event can only be observed and counted once by one observer. One single subject simplifies the philosophical and consequently the mathematical description. In this article the subject is mostly considered to be transcendental and not determined by the laws of nature (exception: end of Section 4.7); but with the further development of chirality theory the subject should be described as a real subset of the universe.

5.3 Universe and Multiverse

A physical observation is a transfer of information from the universe to a subject. Physical means that the observed system is directly or indirectly observable by empirical perception.

In some theories, also a multiverse plays a role:

- The quantum multiverse may be a transcendental set of possible outcomes of measurements or solutions of wave functions.

- It may be a set of individually different universes for every subject.
- There might be other universes which originate from "bubbles" or new black holes in our own universe. Thus, our universe might be the interior of such a bubble or black hole.
- The multiverse may consist of other universes outside of our own.

In chirality theory, there is one and only one universe. This simplifies the theory.

5.4 Perception

The flow of information from the universe to the subject is called empirical perception. The perception might change both the universe as a whole and the subject.

In some theories

- Only part of the universe, namely the perceived objects might be changed by the perception,
- only the universe remains completely unchanged,
- or the transcendental subject remains unchanged,
- or the subject as part of the universe remains unchanged.
- With extra-sensory perception (ESP) the information flow is not empirical.

In chirality theory, both the subject and the observed universe change with every perception. But there might be a transcendental "presumed observer" corresponding to the ego who is not changed by an information transfer. He perceives nature *per se*, i.e. not only real but also other existing elements of nature such as e.g. single points.

5.5 From Language to (Finite) Quantity Value Objects and Category Theory

The physicist describes his observations ("*Anschauung*") of the system S in his daily language $\mathcal{L}(S)$. The meaning of the signals, words or symbols of that language can in principle never be defined precisely. This causes a lot of confusion e.g. when scientists use words such as to be (Be-ing), reality, existence, truth, space, time, nature or measurement. In most articles, such notions are not defined at all.

To apply a theory of type ϕ the physicist then translates the set of observed events $\epsilon_{\phi,S}$ into a mathematical language with quantity value objects $R_{\phi,S}$, i.e. numbers:

$$\epsilon_{\phi,S} \rightarrow R_{\phi,S}.$$

Number language is the most precise mathematical language for physics. But numbers theory is rather abstract. Thus, there are enormous unsolved mathematical difficulties in all modern theories such as quantum theory, loop quantum gravity, spin network theory, (super) string theory, cosmology. One of the most difficult problems is to apply in principle not perceivable mathematical infinities to the always finite perceptions.

In chirality theory, terms such as to be (Be-ing), reality, existence, truth, space, time, nature or measurement are defined. The first representation of the observed events might be a graphical one with arrows as it is used with great success in chemistry. The advantage of such a language is that it is less abstract, but illustrative and easier to understand. The disadvantage is that calculations are not yet possible. Therefore, a procedure is looked for by which the graphs and arrows are further translated into a language with quantity values as it is done in most other theories. This aim is probably best accomplished by category theory with its topos:

$\tau_{\phi}(S)$ is the topos applied to S.

5.6 Order and Chirality

No information transfer, no language and therefore no physical theory can exist without a certain order. Zero order is chaos. This precondition for every physics is often neglected. It is seldom precisely defined what order means, where it comes from and how it is represented. The minimal condition for every order is that there is a duality of things that can be differentiated although they are characterized the same way, e.g. before and after, left and right, clockwise and counterclockwise, positive and negative, yes and no. In most theories, there is simply a time that is thought to exist or even to be real, although never anybody has observed a time *per se*, one in which nothing happens. This (theoretical or real) time has a time arrow as the basis of order. Other theoretical basic notions for the duality are spin and (qu)bit.

In chirality theory, an explicit axiom of order (or chirality) is formulated. In set theory it reads

$$(P_n, P_m) \neq (P_m, P_n).$$

When reading the axiom, the physicist as an observer must know that there is a basic difference between left and right or between before and after respectively. For that the physicist himself has to be chiral – even if he is a transcendental subject. The origin of order in physics is the direction of the information flow from object to subject. This is best expressed by an arrow ("if …, then …") as in category theory.

5.7 Infinity?

In all physical theories, numbers play an important role. In modern theories this is the case for all types of numbers: natural, rational, surds, imaginary, finite and infinite numbers. Thus, number axioms are required for all theories.

For chirality theory, only finite rational numbers are required because all measurements are described by numbers of counted events. Thus, there are only few number axioms, namely the simple Peano axioms for finite numbers. If surds, imaginary or infinite numbers were introduced, e.g. to describe waves or information, this would be part of the mathematical theory and not an aspect of real nature. In chirality theory, there is no infinity.

5.8 Continuum and Separation of the Universe into Objects

Most theories do not describe the universe as a whole but rather a subset called object. The separation of the universe into subsets is always an artificial one and this might falsify the description of a perception. Often the universe is described as a continuum, although it is much more difficult to define precise limits between objects in a continuum than if the universe consists of discrete sets. Also the separation between subject and object is always an artificial one and leads to philosophical problems mainly in quantum theory, even when the observer is not part of the universe.

This is the case also in chirality theory, but this theory should be suitable for the description of the universe as a whole as well as of the perception of single objects such as elementary particles. In chirality theory, there is no continuum at all.

5.9 Background Time and Space

In most theories, time (and often also space) exist as a background for all objects as a prerequisite of the theory. Space and time might be real or not, but they exist. They are usually continuous, although neither space nor time nor a continuum can be observed as such. This way locality plays an important role for most laws of nature, although there are non-local phenomena in physics such as the interaction between entangled particles. But there are a few theories such as spin network theory where the discrete spin network is finite and has no space-time background. But even that theory is local.

In chirality theory, such problems do not arise at all because neither space nor time nor any continuum are prerequisites.

5.10 Local and Non-Local Phenomena

Since 80 years there is much discussion but little agreement about the interpretation of non-local phenomena [Bell (1964)]. A convincing model to describe non-local phenomena is missing [Stanford Encyclopedia of Philosophy (2008)].

In chirality theory information transfer is always local whereas action transfer is always non-local.

5.11 Ontological Real and Ontological Basic Objects

Objects in physics are subsets of the universe, e.g. galaxy super clusters, the sun, man, a measuring apparatus, atoms, photons or neutrinos. A modern definition of the term object may be: An object is a combination of quantities whose present values permit common predictions about these quantities (in the future). Scientists have looked for a basic object for thousands of years. They have suggested water (Thales), the atom (Democritus), elementary particles, space-time loops, strings, spins, qubits or an orthonormal basis of Hilbert space as the ontological basic object. Sometimes it is not clear whether the basis is something real or rather theoretical.

In chirality theory, a system S consists of subsets called objects. The basic ontological object is the point P. It is the smallest possible object. It is one, has no parts, cannot be divided, is identical with itself, has no internal properties and corresponds to a point in topology. The point exists but it cannot be empirically observed; so it is not real and cannot be counted or measured. It seems that a physical theory without such a not observable entity is impossible. Every object S consists of $\{P_n\}$, where $n = 1,2,3, \dots, n \neq \infty$. The model of the real and observable basic ontological object is the 4-point space.

5.12 Natural Laws as Relations between Objects

The relation between different objects is expressed as a natural law. Such laws may define the relative position or state of the objects and the change of these positions and/or states. In most modern theories, this is accomplished by so called fields.

In chirality theory, all points are related to each other. The relation r is expressed by

$$r_{n,m} = (P_n, P_m).$$

In $\{P_n\}$ there are $n(n-1)$ relations r called distance without considering any relations of relations and so on. No fields are used in the model, only points with their relations to one another.

5.13 Superposition and Individuality

In quantum theory, there are superpositions of states with vague ontological status.

In chirality theory, every single point is an individual ontological basic object: $P_n \neq P_m$. There are no superpositions. But the same P can be an element of many real ontological 4-point spaces at the same time and these spaces may overlap.

5.14 Events, States and the Quantum *h*

As a rule, physical theories describe measurements as states of the observed system at a certain time. States may be probabilistic and they stay unchanged for a certain time. Most physicists consider states to be real aspects of nature. The Planck quantum h is the minimal possible angular momentum, the quantum of action, i.e. the measure of any change of a state.

In chirality theory, there are no existing real states. An event is not a state. Only events are observable and real. An event occurs when something changes whereas something other remains unchanged. Time is not required. An event occurs, when any P_m changes its relation to $P_{1,2,\ldots,i}$ by getting in a between relation to any of these other P. In chirality theory, the quantum h is introduced as the measure for the counting of events. If an event does not change the spin of an object, it is a periodic event and no information is transferred to the outside. If the event changes the spin of the object, a quantized information is transferred to the outside. Every such event changes the orientation of the observed object by a rational number $\mathbb{Q}$ of quanta h. This way only rational numbers are required in chirality theory.

5.15 The Arrow of Time

There are whole books written about the arrow of time. Most authors agree that one cause of the arrow is the second law of thermodynamics. But with the interpretation of modern theories such as special relativity theory (objects moving at light speed, twin paradox), quantum electrodynamics (electrons theoretically moving backward in time), quantum chromodynamics (antiparticles) it might be possible that time stands still or even flows backward, at least theoretically. And for the instantaneous (inter)action between entangled particles no time at all is required. The mathematical model for the description of the arrow of time is the set of the real numbers.

In chirality theory, it is the phase arrow ⇒ of the axiom of chirality from which the arrow of time and the chirality of space are derived.

5.16 Elementary Particles

Elementary particles are objects with well-defined intrinsic values which are multiple subsets of the universe. Some elementary particles are stable, others decay spontaneously. A particle may be composed of smaller particles. The terms well-defined, multiple, stable and spontaneously may differ between theories.

In chirality theory, an elementary particle is a (directly or indirectly) perceivable object S_{Pi}, consisting of a definite number of points and a perpetual internal process of periodic events that do not change S_{Pi} and its orientation.

5.17 3-Dimensionality of Classical Space

In most theories, the 3-dimensionality of classical space is an axiom, but C.F. von Weizsäcker with his Ur-theory [Weizsäcker (1985), 379–412] and D.R. Finkelstein with his qubit-formalism with Pauli matrices [Finkelstein (1996), 319–358] independently discovered a theoretical explanation. In these theories chirality is not derived from metaphysics but is introduced mathematically by the imaginary numbers. These are always chiral and have an arbitrary orientation.

In chirality theory particles are only possible as finite periodic processes with 3-dimensionality. At least 4 points in 4 different states are required to produce a periodic event. Such a 4-point space forms a 3-dimensional space. The 4-point space is the real ontological basic object in chirality theory whereas the single point itself exists but is not real. Classical space corresponds to the synthetic *apriori* space of I. Kant, the space required for every "*Anschauung*" [Kant (1781/2003)]. This does not exclude a space of higher dimensionality in nature, but such a higher dimensional space cannot be observed or be a precondition of observation. Thus, in chirality theory there are $2 \times 3 = 6$ additional "concealed" space dimensions due to black mini-holes and black mini-holes within black mini-holes (see Section 4.3).

5.18 Information

Natural laws determine how information is changed in any system. The quantity of information is the number of bits of the system. One bit is a binary alternative of a yes/no-question. In modern physical theories, the quantity of information corresponds to the energy. Sometimes there is not a clear distinction made between the quantity of information of the system, i.e. the number of yes/no-questions, and the content of that information i.e. the answers to the yes/no-questions. According to special relativity theory information cannot travel faster than light. Many physicists think that this must also be the case for the velocity of any action transfer.

In chirality theory information is related to the number of independent orientations in S_{Pi}. With each periodic event of an object in S_{Pi} neither the content of the information, i.e. the orientation of any subset of 4 points, nor the quantity of the information changes. The energy of S_{Pi} remains the same.

But if an event is not periodic, then orientations of some 4-point-spaces are changed and some information and energy is thereby transferred. Every point is an element of all possible 4-point spaces of the universe containing this point, i.e. many of these spaces overlap. Also in chirality theory, information cannot travel faster than light, but the non-local (inter)action without any information transfer is instantaneous or timeless with each event. If some of the 4 points of the 4-point spaces which change their relative orientation are far away from each other, the action according to the axiom of chirality is non-local. According to the axiom of chirality, single points always travel from event to event with light speed. As long as no point has travelled from one subsystem to another one, no information is transferred to the other system. For the action transfer on the other hand this is not required because it is simply a consequence of the timeless change of the configuration and the relative states of 4 points according to the axiom of chirality (Figures 4 and 5). This difference between chirality theory and all the others probably solves the paradox of entangled particles and of the 2-slit-experiment, probably also that of the collapsing wave function. The bit- and qubit-formalisms of other information theories might be translated for their use in chirality theory.

5.19 Space-Time, Duration and Distance

Duration and distance are important parameters of most physical theories. They are the parameters of time, space or space-time. The properties of these entities differ from theory to theory:

- Time and space may ontologically be real, unreal but existing or they are simple ideas.
- Time has one time arrow or two opposite time arrows.
- Space-time is an extended entity or not.
- Space-time is 4-dimensional or it has more than 4 partly concealed dimensions.
- Space-time might be a continuum or it consists of discrete elements.
- In most modern theories space-time is no more Euclidean but curved.
- Movements are either absolute or relative and subjective.
- With Immanuel Kant space is chiral.

Duration Δt is measured with clocks, distance Δl with a one-dimensional measure or as $c \cdot \Delta t = \Delta l$.

In chirality theory, space and time do not exist but they can be theoretically derived from the only entity event. Events always occur when a 4-point space inverts, that is when a point gets between 3 other points of the universe.

In chirality theory, the duration between two moments t_m and t_n of time is a counted number $\mathbb{Q}_{m,n}$ of periodic events. The basic measure is the intrinsic clock of the (real or presumed) observer. Duration is a rational number. If the observed system consists of more than 4 points, more than one 4-point space with its events has to be considered, and duration becomes not a natural but a rational number. The algorithm that determines the counting of the duration in such a system is not yet known. The time between two events in a single 4-point space is the shortest possible time which probably corresponds to the Planck time. It is defined to be 1 (counted event).

In chirality theory, the distance between two observed objects S_m and S_n is the counted number $\mathbb{Q}_{m,n}$ of non-periodic events that occur, when a single point travels by the shortest way from one object to the other. Distance is a rational number. If the observed system consists of more than 4 points, more than one 4-point space with its events has to be considered and distance becomes not a natural but a rational number. The algorithm that determines the counting of the distance in such a system is not yet known.

6 Open Questions

Similarly to some Greek philosophers, more modern ones such as I. Kant and C.F. von Weizsäcker also assumed that the laws of nature are nothing more than the conditions for every perception, and that nature thus becomes a unified whole [Weizsäcker (1971 b)]. "We must always be prepared to change the axiomatic foundation of physics in order to do justice to the facts of observations in the most logically perfect way possible" [Einstein (1970)]. The theory now developed rudimentarily from the axiom of chirality with just four geometric points fulfils the basic metaphysical conditions for a physical theory according to Section 2.10: The metaphysical ego corresponds to the presumed observer. The theory is simple and – at least for me – immediately evident. It describes perceptible things such as, for example, the neutrinos. The events are causally linked, finite and countable. The spin is chiral and supplies the information content of the object as a binary choice, which changes when perceived by a real subject. The theory can already at least intuitively describe a startlingly great number of aspects of modern physics, and even in part deduce them from metaphysics, e.g. the connection between relativity and quantum theory. And all without recourse to concrete physical experiences with the development of the theory. Such experiences served at most as inspiration. I assume that by simple extensions of the outlined theory also the remaining natural constants, all elementary particles with their interactions, indeed thermodynamic and cosmological phenomena as well, can be described [Wehrli (2008)].

This rudimentary description of the strategy of chirality theory must be further developed. Some open questions are:

- Should category theory be applied to chirality theory?
- What happens when there is an unlimited number of points in the universe?
- How can the 4 known physical forces be described by chirality theory?
- How can a real observer be introduced in the chirality theory?
- What is the role of the constants of nature?

- Can a black mini-hole be stable due to its high symmetry?
- Can there be black holes within black holes which exert a specific action on the outside of the black hole? What does this mean for the arrows of time?
- How can the known elementary particles and their strange properties be described by the chirality theory?
- If the universe is interpreted as the interior of a black hole, what is the meaning of big bang, inflation, dark energy and dark matter?

For finding answers to these questions the axiomatic system of the suggested theory is yet to be completed. The theory will never predict what exactly will happen in the future for two reasons. First, the universe is much too complex to be precisely described. Second, since there are things such as the points which must exist but are in principle not real and perceivable, there are actions in our world which remain uncertain or only statistically predictable. Thus, the difference of opinion between Einstein and Bohr concerning the reality and completeness of the quantum theory could be bridged: For Bohr, only the events empirically perceivable by the real observer would be real, whereas for Einstein even the individual points perceivable only by a presumed observer exist. According to this ontology "God really doesn't play dice". The suggested theory with the axiom of chirality thus introduces new perspectives for theoretical physics.

References

Aristotle (1984). Physics. Charlottesville USA: InteLex Past Masters Online Catalogue www.nlx.com.

Bell, J. (1964). On the Einstein-Podolsky-Rosen paradox. Physics I, 3 (pp. 195–200).

Blackmore, S. (2003). Consciousness. London: Hodder & Stoughton.

Boyle, K.R. (1904). Lecture delivered before the Oxford University Junior Scientific Club on May 16, 1893. London: Clay.

Capra, F. (1991). The Tao of Physics: an exploration of the parallels between modern physics and eastern mysticism. New York: Shambhala.

Davies, P. (1989). The New Physics. Cambridge: Cambridge University Press.

Descartes, R. (1641/1960). Meditationes de prima philosophia. In L. Gäbe (Ed.), Meditationen über die Grundlagen der Philosophie 1 (18–19), 16 and 2 (26–27), 23. Hamburg: Meiner.

Döring, A. and Isham, C.J. (2007). I.–IV. A Topos Foundation for Theories of Physics. https://arxiv.org/abs/quant-ph/0703060. Accessed 21 December 2016.

Drieschner, M. (1981). Einführung in die Naturphilosophie (pp. 85–90). Darmstadt: Wissenschaftliche Buchgesellschaft.

Dümont, J. (1999). Formal-ontologische Kategorien in der Mathematik (p. 41). München: Utz.

Einstein, A. and Infeld, L. (1950). Die Evolution der Physik. Wien: Zsolanay.

Einstein, A. cited by C. Seelig (1970). Mein Weltbild. In Seelig, C. (Ed.), Mein Weltbild (pp. 159ff). Frankfurt: Ullstein.

Einstein, A. cited by Wheeler, J.A. (1979). Frontiers of Time. In G. Toraldo di Francia (Ed.), Problems in the Foundation of Physics (p. 431). Varenna Italy: International School of Physics "Enrico Fermi".

Feynman, R.P. (1965). The Character of Physical Law. Cambridge USA: MIT Press.

Feynman, R.P. (1966 a). Lectures on Physics (Vol. III, 1–10). Reading: Addison-Wesley Publ. Co.

Feynman, R.P. (1966 b). Quantum Mechanics (p. 17–7). Reading: Addison-Wesley Publ.Co.

Feynman, R.P. (1985). The Strange Theory of Light and Matter. Princeton: Princeton University Press.

Finkelstein, D.R. (1996). Quantum Relativity. A Synthesis of the Ideas of Einstein and Heisenberg. Berlin: Springer.

Greene, B. (1999). The Elegant Universe. Superstrings, Hidden Dimensions, and the Quest for the Ultimate Theory. New York: Norton.

Grünbaum, A. (1973). Philosophical Problems of Space and Time. Boston: Reidel.

Hampe, M. (1998). Alfred North Whitehead (p. 31). München: Beck.

Hilbert, D. (1926). Über das Unendliche. Mathematische Annalen 95, pp. 161–190.

Hofstadter, D. (1985). Gödel, Escher, Bach (pp. 416, 755f). Stuttgart: Klett.

Kant, I. (1768/1960). Vom ersten Grunde des Unterschiedes der Gegenden im Raum. In W. Weischedel (Ed.), Kant Werke in zwölf Bänden (pp. 991–1000). Wiesbaden: Insel.

Kant, I. (1781/2003). Kritik der reinen Vernunft. J. Timmermann (Ed.) A154–A166 (pp. 253–264). Hamburg: Felix Meiner.

Kant, I. (1783/2001). Prolegomena zu einer jeden zukünftigen Metaphysik (§ 36, 39). Hamburg: Meiner.

Kant, I. (1786/1957). Kritik der Urteilskraft (p. 16). Wiesbaden: Insel.

Kuhn, T. (1967). The structure of scientific revolution. Chicago: Chicago University Press.

Leibniz, G.W. and Clarke, S. (1715–16). In E. Dellian (Ed.), Der Briefwechsel zwischen Leibniz und Clarke (1990). Hamburg: Felix Meiner.

Mac Lane, S. (1971). Categories for the Working Mathematician. New York: Springer.

Mac Lane, S. (1998). Categories for the Working Mathematician (pp. 31f). New York: Springer.

McDonald, A.B. et al. (2003). Solving the Solar Neutrino Problem. Scientific American April 2003, pp. 22–30.

McLarty, C. (1992). Elementary Categories, Elementary Toposes. Oxford: Clarendon Press.

Penrose, R. (2004). The Road to Reality. London: Cape.

Quine, W.V.O. (1963). Set Theory and its Logic (pp. X, 9ff, 241ff). Cambridge: Harvard University Press.

Rovelli, C. (2010). Quantum Gravity. Cambridge: Cambridge University Press.

Russell, B. (1903/1956). The Principles of Mathematics (pp. 199-217). London: Allen & Unwin.

Shapiro, S. (1997). Philosophy of Mathematics. Structure and Ontology. Oxford: Oxford University Press.

Smolin, L. (2001). Three Roads to Quantum Gravity. London: Phoenix.

Stanford Encyclopedia of Philosophy (2008). https://plato.stanford.edu/archives/win2008/entries/qm-action-distance/

Suppes, P. (1972). Independence of the Axiom of Choice and the Generalized Continuum Hypothesis. In P. Suppes (Ed.), Axiomatic Set Theory (pp. 252f). New York: Dover.

’t Hooft, G. (2015). The Cellular Automaton Interpretation of Quantum Mechanics. https://arxiv.org/abs/1405.1548. Accessed 21 December 2016.

Wehrli, H. (2008). Metaphysics. Chirality as the Basic Principle of Physics. www.hanswehrli.ch.

Weinberg, S. (1999). A Unified Physics by 2050? Scientific American December 1999 (p. 36).

Weizsäcker, C.F. von (1971 a). Voraussetzungen naturwissenschaftlichen Denkens. Die Unendlichkeit der Welt (pp. 53–87). München: Hanser.

Weizsäcker, C.F. von (1971 b). Die Einheit der Natur. München: Hanser.

Weizsäcker, C.F. von (1985). Aufbau der Physik. München: Hanser.

Weizsäcker, C.F. von (1999). Grosse Physiker (pp. 344–348). München: Hanser.

Whitehead, A.N. (1936). Process and Reality. An Essay on Cosmology. New York: Macmillan.

Whitehead, A.N. (1939). Science and the Modern World (pp. 103ff). New York: Macmillan.

Wolschin, G. (1999). Kaonen spüren den Zeitpfeil. Spektrum der Wissenschaft April 1999 (pp. 14ff).

Zermelo, E. (1908). Untersuchungen über die Grundlagen der Mengenlehre I. Mathematische Annalen 65 (pp. 261–281).

FSC
www.fsc.org
MIX
Papier | Fördert
gute Waldnutzung
FSC® C083411

Zeitfracht Medien GmbH
Ferdinand-Jühlke-Straße 7
99095 Erfurt, Deutschland
produktsicherheit@kolibri360.de